Volcanos in the Atlantic:
4. Iceland
5. Canary Islands

Volcanos in Europe:
6. Southern Italy

Volcanos in Asia:
Few large volcanos

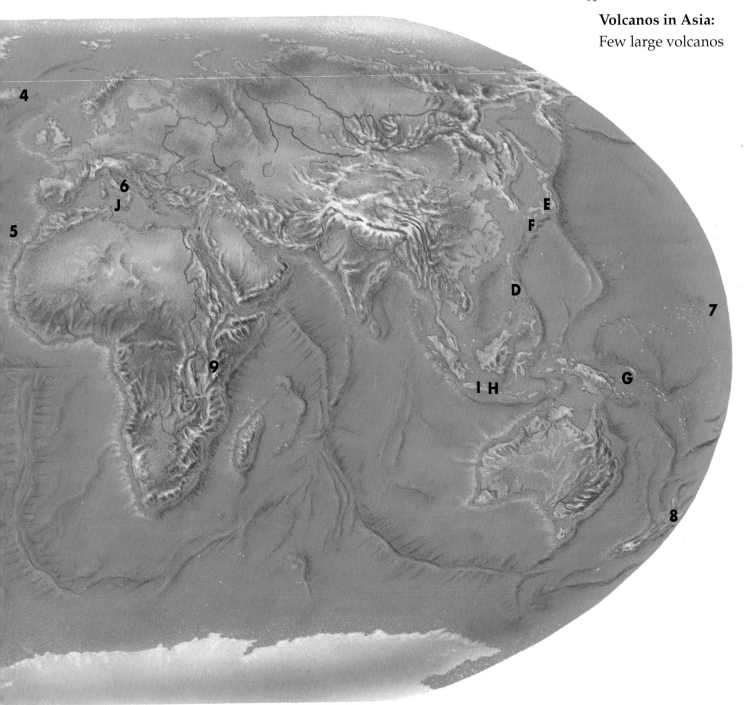

Volcanos in Australasia:
8. North Island, New Zealand

Volcanos in the Pacific:
7. Almost all Pacific islands
   are volcanos

3

FACTS ABOUT VOLCANOS

There are nearly 900 active volcanos in the world, most of them situated in a line that stretches around the edge of the Pacific Ocean. This line is known as the Pacific Ring of Fire.

Perhaps the most famous historic eruption was Mt Vesuvius in Italy in AD 49. Hot gases killed those people who had remained in the town of Pompeii, while a mud flow killed people in nearby Herculanium.

The largest eruption within the last 2000 years was in New Zealand when Mt Taupo erupted in AD 130, ejecting 30 billion tons of rock and flattening over 5,000 square miles of the surrounding land.

In the eruption of Tambora, a volcano in Indonesia, in 1815, the total amount of material shot into the air was over 36 cubic miles. When it was over, the top 5000 ft of the volcano had been blown away.

The eruption of the volcano on the Indonesian island of Krakatoa in 1883 was so powerful that the sound was heard over a third of the Earth's surface, making a sound like the roar of heavy guns.

The world's tallest land volcano is Cerro Aconcagua in the Peruvian Andes, nearly 23,000 ft high, but the world's tallest active volcano is Mauna Kea and the world's largest active volcano is nearby Mauna Loa, both part of the volcanic Hawaiian islands chain in the Pacific Ocean.

The most famous recent volcano to erupt is Mt St Helens in Washington State, USA. It erupted under the full glare of the world's TV cameras on 18th May 1980. However dramatic the pictures of it may seem, it was a baby volcano in the world scale of eruptions. There has been at least one significant volcanic eruption each year since then.

The largest eruption this century was Mt Pinatubo in the Philippines which erupted on 15th June 1991. The ash and dust thrown into the air from such a volcano makes the world's weather slightly cooler for several years afterwards.

Grolier Educational Corporation
SHERMAN TURNPIKE, DANBURY, CONNECTICUT 06816

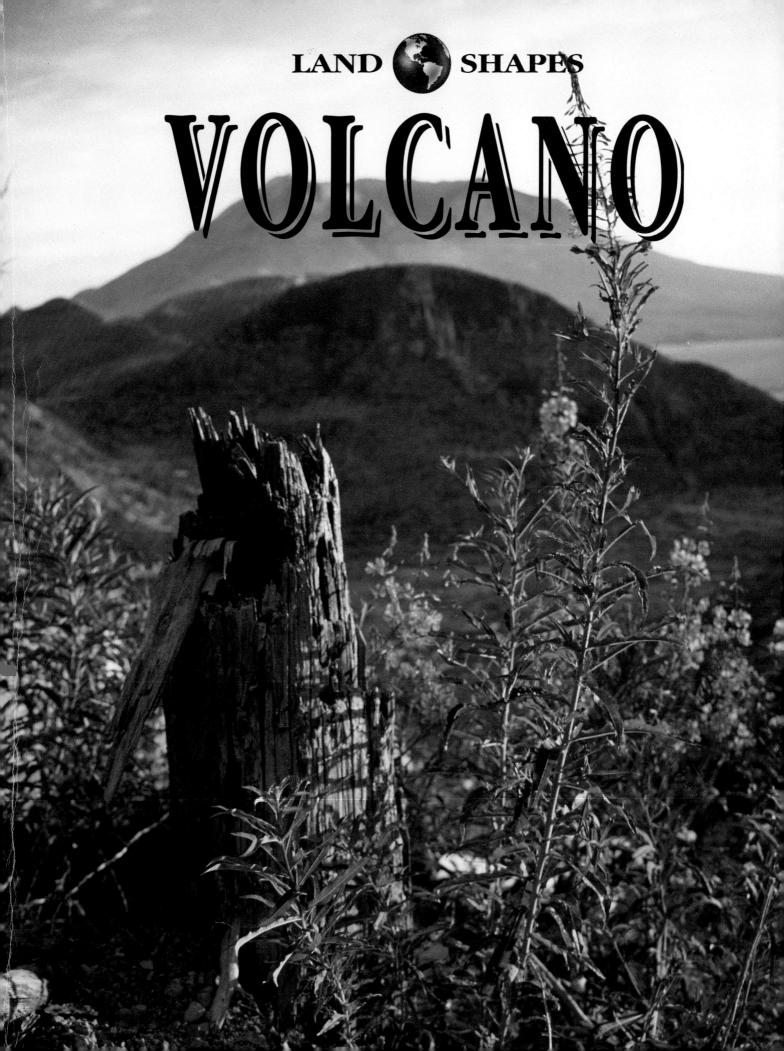

LAND SHAPES
VOLCANO

Author
Brian Knapp, BSc, PhD
Art Director
Duncan McCrae, BSc
Editor
Rita Owen
Illustrators
David Hardy and David Woodroffe
Print consultants
Landmark Production Consultants Ltd
Printed and bound in Hong Kong
Produced by
EARTHSCAPE EDITIONS

First published in the USA in 1993 by
GROLIER EDUCATIONAL CORPORATION,
Sherman Turnpike, Danbury, CT 06816

Copyright © 1992
Atlantic Europe Publishing Company Limited

Library of Congress #92–072045

Cataloging information may be obtained
directly from Grolier Educational Corporation

Title ISBN   0–7172–7185–4

Set ISBN   0–7172–7176–5

Acknowledgements. The publishers would like
to thank the following: Redlands County Primary.

Picture credits. All photographs from the
Earthscape Editions photographic library except
the following (t=top, b=bottom, l=left, r=right):
Austin Post/USGS 20b, 30l; Keith Ronnholm 12;
NASA 34b; Stephen Porter 31t; USGS 13b, 18bl, 34/
35, 35t; ZEFA *Cover*, 8/9, 10/11, 15t, 22/23, 24l, 31.

Cover picture and inside back cover picture:
Mt St Helens, Washington, USA.

In this book you will find some words that have been shown in **bold** type. There is a full explanation of each of these words on page 36.

On many pages you will find experiments that you might like to try for yourself. They have been put in a blue box like this.

In this book mi means miles and ft means feet.

These people appear on a number of pages to help you to know the size of some landshapes.

CONTENTS

World Map   2

Facts about volcanos   3

Introduction   8

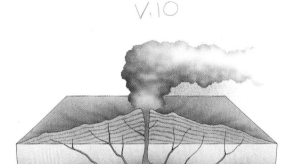

Chapter 1: The world of volcanos

What volcanos do   10

Where volcanos erupt   12

The nature of eruptions   14

Chapter 2: Lava

What is lava?   16

Rivers of fire   18

Chapter 3: Ash and gas

Ash, bombs and explosions   20

Building with ash   22

Chapter 4: Cones and craters

The shapes of cones   24

Craters   26

Ancient volcanos   28

Chapter 5: Volcanos of the world

Mt St Helens   30

Hawaiian islands   34

New words   36

Index   37

Introduction

When a volcano **erupts** it is sure to become world news because volcanic eruptions are some of the world's most spectacular natural events. Volcanos are also often unpredictable and can be extremely violent.

A volcano hints at how thin and fragile the Earth's **crust** really is. Under the world's oceans it can be as little as 4 mi thick and under land sometimes no more than 24 mi. Below the crust, however, there is a different world, one of searing heat, trapped gases and liquids ready to burst forth wherever there is a weak place in the crust.

Although the world seems to be dotted with volcanos, weak places in the crust are more common in some places than others. Hence volcanos may be common in some countries and absent from others. Scientists have found that volcanos vary enormously. In some areas volcanos can produce terrifying explosions, sending great clouds of gas and **ash** high into the air, yet in other places they are relatively quiet and produce **lava** that you can even walk away from.

Much of the Earth's crust was made from ancient volcanos. Volcanos are still building new land today. Find out about the spectacular landshapes that are produced and how to look for signs of ancient volcanos near to your home. Simply turn to a page and discover the landshapes of volcanos.

Chapter 1:
The world of volcanos

What volcanos do

Volcanos are the builders of the world's land. Nearly all the rocks found in the world were first laid down on the Earth's surface as lava and ash.

There are many volcanos that are still active and which erupt every few hundred years or so. Some are said to be **dormant** because scientists think that they will erupt in time; others seem to have finished bringing molten rock to the surface and are called **extinct** volcanos.

All volcanos make large landshapes, and are among the easiest land features to spot.

Shapes and materials

Many volcanos build a cone near to where they erupt, while others produce lava that spreads out in vast level sheets.

Volcanic cones are a mixture of the fine ash that settles out of the air and the streams of lava that flow directly from the crater. The exact nature of the material will eventually control the shape of the **crater**.

Many volcanos send large amounts of ash high into the sky (see page 20).

Ash and lava build up to make a cone (see pages 16 and 24).

Lava often flows from a volcano in a narrow channel (see page 18).

The volcano is erupting a fountain of molten lava (see page 16).

The pit in the top of the cone is called the crater (see page 26).

Where volcanos erupt

Volcanos come in all shapes and sizes. Although they are scattered throughout the world there is a pattern to where they are found, and how explosively they erupt.

(Nearly all volcanos are found in regions where the Earth's crust is very weak, making it easy for molten rock, known as **magma**, to force its way to the surface. These places are the boundaries to huge slabs called **plates** that make up the Earth's crust.)

Violent outbursts

Some volcanos erupt with very little warning, sending huge clouds of steam and rock fragments called ash into the air. Many of them mark the places where two plates are pushing together.

As ash and gas rush out the side of the volcano is blasted away.

The land nearby is covered with thick layers of ash.

Summit of volcano.

Plates

This picture shows the most important cracks in the Earth's crust and the main plates that lie between the cracks. Volcanos are especially common at the boundaries.

Volcanos that form at this boundary under the sea are usually quiet and mostly erupt lava.

B

A

The volcanos that erupt at this boundary are usually explosive.

An earth plate contains both continent and ocean floor. This is South America and also part of the Atlantic Ocean floor.

This is the African plate. The part of it above the sea we know as Africa.

This volcano is jetting up fountains of lava.

B

As the lava flows away it will build a broad cone.

The nature of an eruption

Volcanos erupt as the build up of molten rock deep inside the Earth breaks through the crust.

We cannot see directly how a volcano works, but by looking at a furnace for making iron we can get an idea of some of the things that might be happening.

The source of lava and ash

Volcanos are fed from chambers filled with molten rock (magma) deep in the crust. Volcanic eruptions can only go on for as long as there is magma under high pressure in the chamber.

An eruption usually drains the chamber quickly. Then the volcano becomes quiet, lava solidifies in the pipe and the volcano becomes sealed up.

Over perhaps many hundreds of years, the chamber refills and pressure increases. In time the volcano will erupt once more.

The magma forces its way through the Earth's crust along lines of weakness. Where it reaches the surface ash and lava are produced.

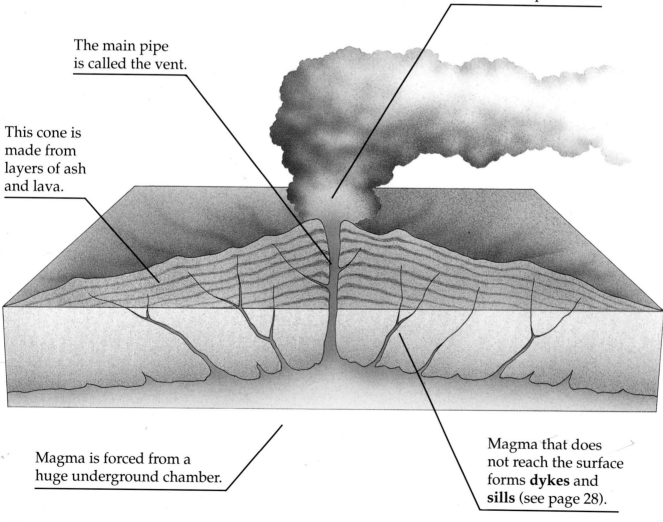

The main pipe is called the vent.

This cone is made from layers of ash and lava.

Magma is forced from a huge underground chamber.

Magma that does not reach the surface forms **dykes** and **sills** (see page 28).

3. The molten rock contains many gases under great pressure. During an eruption they are emitted as towering clouds along with small pieces of lava that cool to make ash.

Huge clouds of steam and other gases are also released from the top of a blast furnace.

The waste rock (called slag) and the iron are released from low down in the furnace. As they leave the furnace they are yellow or orange hot, similar to the lava in a volcano.

2. Molten lava flows out from the volcano beneath the ash cloud. At first it is so hot it is bright yellow. As it cools it becomes orange and finally black.

The purpose of an iron furnace is to make the iron ore so hot it will become molten.

1. The rocks that will be the source of the eruption are heated deep within the Earth.

Chapter 2:
Lava

What is lava?

Lava is the name given to both the liquid that flows from some volcanos and the rock that forms as the liquid cools.

There are many different types of lava, some are runny like water and others are sticky like molasses. Sometimes lava flows out of cones and **fissures** on land, but much more flows unseen across the ocean beds. Each type of flow makes its own distinctive lava.

Pumice is the world's only floating rock! It floats because it contains many bubbles, made by gases inside the lava trying to escape. However, the lava cooled before the bubbles could get out, trapping them for ever.

SCALE

This is ropy or **pahoehoe** lava. You can see the ropy shapes in this piece.

In the main picture below you can see the edge of a rough and broken lava flow. This type of lava is called **aa-aa**. It became like this just before it stopped moving.

Lava flows easily when it is very hot, but as it flows across the ground it is cooled against the air and it develops a crust. This lava has a crust, but you can see that it is still molten within.

Rivers of fire

Basalt is a black-colored lava. When it is hot, basalt is the most runny of all lavas, it can flow as fast as a speedboat for hundreds of miles from its home crater or fissure.

Most of the time it erupts quietly and forms modestly-sized sheets that may cover a few hundred square miles. But just occasionally the crust opens wide and vast amounts of lava flood out over the land.

Imagine an area of searing hot lava the size of the North American Great Lakes, and you get an idea of how big some basalt flows in the past must have been.

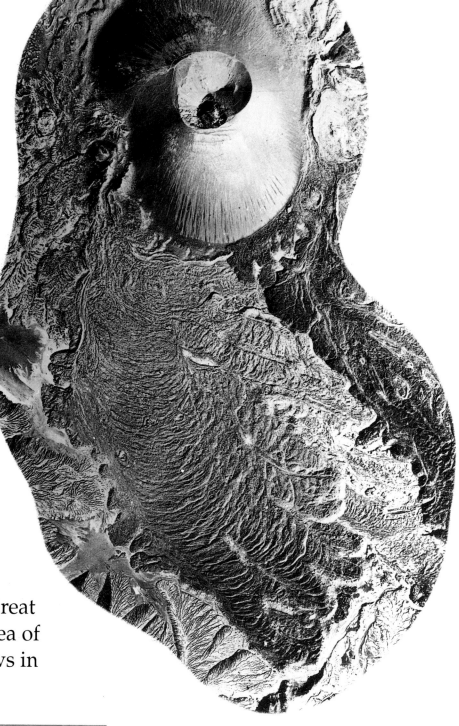

The picture above was taken from high above a crater. You can see old lava flows at the base of the crater. There are many different ages of lava flow, and you can see one flow overlapping another. In this way huge thick sheets of lava can cover a landscape.

The edges of the flow seen in the picture on the left have cooled and become solid, trapping a single 'river' of lava. By keeping in a narrow channel the lava holds the heat and can travel great distances.

The Columbia River area of north western USA was once covered in flood basalt. Here it can be seen making the cliffs of the Columbia River Gorge.

Fissure eruptions

Sometimes the Earth's crust opens along a huge fissure. This kind of 'wound' can bleed lava for weeks, months or even years, sending out huge floods of lava that can bury the surrounding countryside.

Eventually the basalt makes sheets of rock that can be traced for hundreds of miles.

A piece of a basalt column.

These basalt columns are lying on their side. They are part of a flood of basalt lava that slowly cooled many millions of years ago.

Chapter 3:
Ash and gas

Ash, bombs and explosions

Ash is the fine material thrown high out of a volcano when it erupts. The ash can be pieces of rock broken from the pipe walls or it may be tiny pieces of lava that have cooled while in the air. Ash only forms when a volcano explodes.

This is a volcanic bomb, a large piece of lava that cooled and became solid as it flew through the air.

Why explosions happen

Molten material (magma) deep beneath a volcano is under great pressure. As it erupts, the pressure on this material lessens, and many substances immediately swell and become gases. This rapid change causes an explosion, hurling gas and tiny droplets of molten rock high into the air, creating clouds that tower over a volcano.

At the start of an eruption, boulders weighing over 200 tons may be thrown many miles from the volcano. This is one reason that standing within sight of an erupting volcano is not a very safe thing to do! The picture below shows Mt St Helens at the start of the 1980 eruption.

See how explosions happen

A bottle of drink is made fizzy in a factory by dissolving gases in it and then putting the cap on before the gases have a chance to escape. In some ways this is similar to the liquid trapped below the Earth's crust.

If you shake a fizzy drink bottle hard and then remove the cap, the drink will spurt out as the gases expand to make bubbles.

Even when most of the gases have been released, the drink will continue to bubble over for a while longer.

This is also the way a volcano erupts. First a huge cloud of gases are thrown out of the pipe, and later lava wells up to the surface and spills down the sides of the cone.

When the violent fizzing has stopped look to see how much liquid has flowed out of the bottle. If you think of the bottle as the chamber where the liquid rock gathers deep in the crust, you can see how an eruption quickly empties part of the chamber.

Part emptying leaves the roof of the chamber unsupported. Above lies the whole weight of the volcano's cone. Sometimes the roof collapses, causing the volcano to collapse. This is what has often happened when the top of the volcano appears to have been blown away.

When Mt St Helens erupted in 1980 it added to the land as well as blowing the top of the mountain away. Over a fifth of a cubic mile of new rock was belched from deep within the Earth. Lumps of cinder settled near to the cone, but fine ash was spread for many hundreds of miles.

Some of the ash was later melted down and made into small trays as souvenirs. The cinders and ash are mostly made of the same mineral (called quartz) that makes glass! In the picture above you can see cinders and ash inside a glass souvenir tray.

The power of a blast to change the landscape can be imagined from this picture. This is what happened to a car which was over 5 mi away when Mt St Helens exploded.

Building with ash

When volcanos erupt they often send a towering plume of ash and gas high into the sky. Most of the ash lands near to the vent and helps to build the volcano's cone.

Witnesses to landshape building

The people of this town of Heimaey in Iceland became unwilling witnesses to an eruption on their doorsteps. The eruption sent large quantities of ash into the air which then landed on their houses and in the streets.

Can you see from this picture how the houses that are closest to the volcano have the greatest amount of ash on them? Can you imagine what problems this creates and why many people have already scraped the ash from their roofs even though the volcano is still erupting?

Chapter 4:
Cones and craters

The shapes of cones

The most easily noticed landshape of a volcanic eruption is the cone of a volcano. The shape of the cone also tells us a lot about the type of volcano it is. Here are some clues to cone-spotting.

Mt Fuji in Japan is a 'classically' shaped cone made up of thousands of layers of ash and lava. The center of the volcano is made up of flows of lava and coarser ash. That is what makes it quite steep-sided. Further away from the pipe the sides of the cone are more gentle because it is built from finer ash.

A cone builds

A volcanic cone is made up from the lava, ash, cinders and bombs that are ejected from the pipe.

The rule for cone-spotting is: the runnier the lava the broader the cone, while the stickier the lava the steeper and smaller it is. Ash always builds to form steep-sided cones, but the steepest cones of all are made from cinders.

The work of lava

The lava builds up as flows. Each new flow will move in a different direction down the side of the cone, gradually building up evenly all around the mountain.

Make lava flow

To see why it is difficult to predict how an eruption of lava will change the land, put some fairly runny Plaster of Paris in a cloth used for icing a cake. Tie it to a tube buried in a cone of plaster.

Squeeze the bag sharply and some plaster will spurt over the cone, running down the side like a lava flow. Now guess where the next 'lava flow' will go, then squeeze the bag sharply again.

In this way you can see how a cone is formed from many different eruptions.

The small cones in the picture below are largely made from golf-ball sized cinders.

Craters

As the volcano erupts, the huge amounts of fine rock and gases that are blasted from the pipe act like powerful sandpaper, scouring the supply pipe or vent inside a tube.

At the top of the pipe there is little rock to resist the blast, and the upper part of the cone is often blasted away to make an irregular shape depression, or crater. After the eruption has ceased the sides of the crater slump in and make the crater cone-shaped.

Craters have no natural outlets for any rain that falls and many soon fill up with water to make crater lakes. The world's most famous volcanic lake is Crater Lake, Oregon, USA. It is 7 mi across and the crater over 3000 ft deep.

Notice that a new cone, called Wizard Island, is building inside the old one and it has now grown above the level of the lake.

For more information on crater lakes and collapsed volcanos see the book 'Lake' in the Landshapes set.

Collapsed sides of the cone.

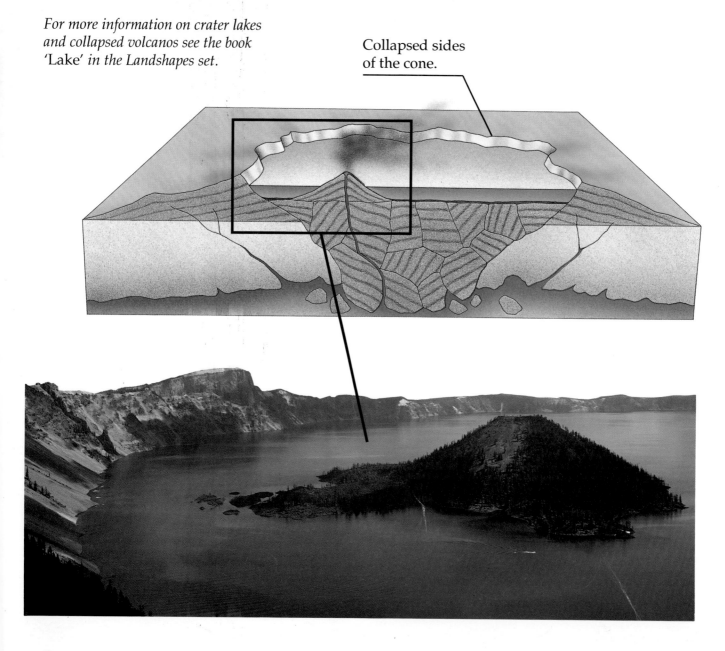

This view of
a crater which
was recently
active shows
the steep-sided
cone leading
down to the
central pipe.
The pipe has been
sealed with solid
lava and only wisps
of steam can be seen.

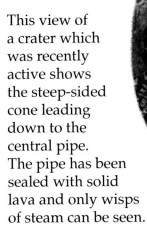

Make a crater
To see how a crater
works, fit a length of
plastic tube to the end
of a bicycle pump and
bury it inside a mound
of dry sand.

When you use the
pump the pressure will
cause the sand to fly
out just as when a
volcano erupts. Look at
the resulting crater and
compare the sides with
those shown in the
pictures on this page.

This picture below shows several craters on Mt Agung,
Indonesia which erupts every few years. The weather
does not have time to work on the cone between
eruptions and so the cone looks sharp and fresh.

Ancient volcanos

An eruption may last for a few days, a few months, or many years. But sooner or later the pressure that caused the eruption lessens until there is not enough force to expel the lava out of the pipe. Eventually lava will cool and turn into rock inside the pipe, making a plug which can be thousands of feet thick and which will seal the volcano until it is blown apart by the next eruption.

Ancient plug
A plug of lava makes a tough pillar of rock that will stand up to the weather when the rest of the cone has been **eroded**.

Some plugs are the remains of giant volcanos and they are even big enough for entire towns to be built on them.

Even the smaller plugs have been used as the sites for ancient churches.

The central pipe is not the only route that lava may follow to the surface. Many volcanos break through their cones and send side pipes to the surface.

Side pipes eventually fill with lava just like the central cone. They are made of the same tough rock and you can see them on the diagram on page 14. You can also see an example at the base of this plug in Arizona, USA, where they look like fins supporting the base of a rocket. Each of the fins is called a dyke.

Shaded area shows where the original volcano would have been.

SCALE

This is an area of granite moorland in Dartmoor, UK.

Granite domes

When a volcano becomes extinct, the magma chamber that fed molten rock when it was active slowly cools and turns to a solid rock called granite.

The granite of a magma chamber is only seen at the surface when thousands of feet of rock have been eroded away. A magma chamber is often shaped like a large volley ball, and its broad flanks can sometimes be seen as rounded mountains or moorlands. If you find granite rocks underfoot, then you are walking over the magma chamber of an ancient volcano!

Le Puy, France, a small plug with a beautiful rod-like shape.

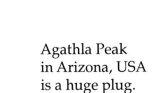

Dyke

Agathla Peak in Arizona, USA is a huge plug.

Chapter 5:
Volcanos of the world

Mt St Helens

This volcano erupted spectacularly on the morning of 18th May 1980. Before this it had been dormant for 150 years.

Scientists knew that the volcano was about to erupt because for some months it had been causing earthquakes.

For a few weeks before the final eruption a large bulge grew and grew, showing where the eruption was going to happen. Then, just after 8 o'clock in the morning, the rocks of the bulge fell away and a huge blast tore the mountain top apart.

The top of the volcano has been blown away. The place where the main sideways blast occurred is clearly seen as a deep gash in the rim of the crater.

Instant change
The eruption of Mt St Helens shows just how rapidly volcanos can change. In just a few minutes the top of the volcano had disappeared and a huge crater was formed; land up to 20 mi away was affected by the blast and ash was scattered over an area of thousands of square miles.

Before Mt St Helens erupted it was thought to be one of the most beautiful volcanos in the Cascade Range of northwestern USA. In the distance to the left is Mt Ranier, and to the right, Mt Adams.

This lake (Spirit Lake) has been partly filled with trees that were snapped like matchsticks by the blast.

This area used to be completely covered with forests. Now there are no trees.

Ash fell over the entire area you can see in this picture.

Mt St Helens: Ashes to ashes, dust to dust

One of the main effects of the eruption was to send huge amount of ash right across America. The local radio station included this announcement:

'If you are thinking of spending your vacation at Mt St Helens this summer don't bother: Mt St Helens is coming to visit you instead!'

The land had been changed so much within a few minutes that it was impossible to recognise the area around the volcano as it was before. This area was a forest just a few hours before the picture was taken. The eruption knocked over the trees, then covered the land with ash.

This is the profile of Mt St Helens today.

Walking through the ash and fallen trees after the eruption.

Within minutes of the start of the eruption, the ice cap that had covered the summit of Mt St Helens had melted and torrents of water and debris were rushing down the mountainside.

After a few hours many valley bottoms were filled with a thick layer of mud. (Each of the trees in this picture is about 200 ft long.)

The day after the eruption the area around Mt St Helens looked as though it would never recover. Yet, as this picture shows, within 10 years plants were growing back.
The ash has already started to change into soil. Soon it will be hard to tell that the eruption ever happened.

Hawaiian islands

The Hawaiian islands make a chain of volcanos that stretch over 600 miles of the Pacific Ocean. But these volcanos are very different from the explosive volcano of Mt St Helens. Instead of throwing out ash, the Hawaiian volcanos pour streams of lava from their craters. This makes them much safer landshapes to live near than Mt St Helens. You can even visit an active volcano because one area has been made into a National Park.

The black basalt lava spreads out to form huge domes that rise from the sea bed. The islands that people live on are just the tiny tips of the volcanos near the crater. These volcanos rise over 30,000 ft from the ocean floor.

Basalt is easily eroded by running water, which is why the sides of the volcanos are scored with numerous valleys.

This picture shows one of the lava flows from the air.

Flooding lava
When the Hawaiian volcanos erupt they send out huge rivers of molten lava. In the back of this picture you can see a fountain of lava being sent into the air. This is the source of the 'waterfall' of lava that makes up the nearer part of the picture. The lava is over 1800 ℉!

New words

aa-aa lava

this is a kind of lava that is quite sticky and which, as it cools, moves with difficulty. Its surface tends to break up into piles of broken rock

ash

the very fine material that falls from clouds above a volcano. Much of the ash would have started out as fine spatters of liquid lava which cooled in the air. The rest would be fragments of shattered rock

crater

the depression that forms in the center of a volcano. Sometimes craters fill with water to give crystal-clear lakes

crust

the name for the layers of rock that form a solid skin to the surface of the Earth

dormant

a volcano which has not been active for many years, but which scientists think will erupt in the future. Some volcanos may only erupt once every thousand years. It is quite difficult to know whether a volcano is dormant or extinct

dyke

a sheet of lava that once cut across layers of rocks and then turned into a solid. Dykes are found as ridges in the landscapes in many countries. They show that volcanos were once active in the area

erode

the breakdown and carrying away of rocks by water, ice and wind

erupt

the pattern of events that happen when a volcano becomes active. First it causes earthquakes, then the vent is blown clear and finally ash and/or lava come out of the vent

extinct

the name given to a volcano that scientists think has stopped erupting for ever

fissure

the name for a deep crack in the crust that may go all the way to the magma deep below the Earth's surface. When lava cools in a fissure it makes a dyke

lava

the liquid material that flows from a volcano

magma

molten material that flows from deep within the Earth. It is the source of all materials that come from a volcano during an eruption

pahoehoe lava

the runny kind of lava that flows for quite long distances before it cools and hardens. It often forms rope-like patterns on its surface

plate

large areas of the Earth's crust. The crust is made up of a number of plates which are constantly moving

sill

a sheet of lava that forced its way in between other layers of rock

vent

the central pipe of a volcano leading from the magma chamber to the surface

Index

aa-aa lava   17, 36
Africa   13
Agathla Peak   28
Andes   4
ash   8, 10, 12, 20, 21,
     23, 30, 33, 36
Atlantic Ocean   13

basalt   18, 19
basalt column   19
bomb   24

Cascade Range   31
Cerro Aconcagua   4
chamber   14, 21
cinder   24, 25
Columbia River   19
Columbia River Gorge   19
cone   10, 13, 24, 25
continent   13
crater   10, 18, 26, 27,
     30, 36
Crater Lake   26
crust   8, 14, 18, 36

Dartmoor   29
dormant   10, 36
dyke   14, 28, 36

earthquake   30
erode   28, 36
erupt   8, 20, 28, 36
explosive   13
extinct   10, 36

fissure   16, 19, 36
France   29

gas   8, 21
granite   29.

Hawaiian islands   34
Heimaey   23
Herculanium   4

Iceland   23
Indonesia   4, 27
Italy   4

Japan   24

Krakatoa   4

lava   8, 10, 16, 20, 21, 36
lava flow   17, 18
Le Puy   29

magma   12, 14, 36
Mauna Kea   4
Mauna Loa   4
Mt Adams   31
Mt Agung   27
Mt Pinatubo   4
Mt Ranier   27, 31
Mt St Helens   4, 21,
     31, 32
Mt Taupo   4

ocean   8
ocean floor   13
Oregon   26

Pacific Ocean   34
Pacific Ring of Fire   4
pahoehoe lava   17, 36
Peru   4
Philippines   4
pipe   14, 26, 27, 28
plate   12, 13, 36
plug   28
Pompeii   4
pressure   20
pumice   15

ropy lava   17

sill   14, 36
slag   15
South America   13
Spirit Lake   31
steam   12, 15

Tambora   4

vent   26, 36
volcanic bomb   20
volcano   8

Washington State   4
Wizard Island   26

MAR 1 ▥ 1994

MAR 1 ▥ 1994